WILLIAMS' PERRAN FOUNDRY CO.

Illustrated catalogue of pumping
and winding engines   090404002X

ILLUSTRATED

OF

Pumping & Wi...

AND OTHER PLANT USED F...

HORIZONTAL, FIXED, AND

CORNISH AND OT...

GENERAL MA...

MANUFACTUR...

WILLIAMS' PERRAN FOUNDRY CO.,

PERRANARWORTHAL, CORNWALL,

AND

1 & 2 GREAT WINCHESTER STRE...

LONDON, E.C.

ESTABLISHED 1795.

NOTE 1.—In addition to the contents of this Catalogue, **WILLIAMS' PERRAN FOUNDRY CO.** manufacture, to **Engineers' Drawings** and **Specifications**, every description of **Heavy Fixed Powers, Sugar Plant, Castings** and **Forgings, Machinery** for **Water, Sewage,** and **Gas Works,** and all kinds of **Machinery** and **Tools** for **Mining Purposes,** &c.

NOTE 2.—**WILLIAMS' PERRAN FOUNDRY CO.** request their **Correspondents,** especially their **Foreign Correspondents,** to give with their inquiries the fullest data possible, so as to leave no doubt of their wishes and requirements.

LETTS, SON AND CO., PRINTERS, NEW CROSS, S.E.,
AND
8 ROYAL EXCHANGE, E.C.

WILLIAMS' PERRAN FOUNDRY CO. have not prepared this Catalogue with a view to give detailed information of the cost and powers of the Machinery they manufacture. In Mining and Pumping operations, which vary in detail, and work to be done, with almost every situation, it would be very difficult to publish Tables of Prices.

Their intention is simply to give those Engineers and others, both abroad and at home, with whom they may not yet have done business, some little idea of the class of Machinery they supply, and which an experience of three-quarters of a century has enabled them to bring to its present great perfection for every country where such powers are employed.

WILLIAMS' PERRAN FOUNDRY CO. will at all times be happy to forward, to any part of the world, Estimates, with Plans and Specifications, in accordance with inquiries, or will furnish prices to Engineers' designs.

1 & 2 GREAT WINCHESTER STREET BUILDINGS,
LONDON, E.C.

LA COMPAGNIE DES FONDERIES WILLIAMS PERRAN, na'pas prèparé ce Catalogue dans le but de donner des renseignements détaillés, sur le prix et la force des machines qu'elle fabrique. Les opérations de creusement et puisement des mines varient dans leurs détails et dans le travail à exécuter, presqu' à chaque dégré de profondeur, et il serait très difficile de publier une table des prix.

Son intention est simplement de donner aux Ingénieurs et au public, tant en Angleterre qu' à l'ètranger, avec qui elle n'a point encore fait d'affaires, un apperçu du genre des machines qu'elle fournit, et qu'une expèrience des trois quarts d'un siècle a'mise a même de porter à leur perfection actuelle, ce qui les rendent précieuses pour les pays òu elles sont employées.

LA COMPAGNIE DES FONDERIES WILLIAMS' PERRAN se fera un plaisir en tout temps, d'envoyer dans toutes les parties du monde des devis, avec plans et spécifications conformes aux demandes qui lui seront adressées, ou ses prix d'après les dessins des Ingénieurs.

Nos. 1 & 2 GREAT WINCHESTER STREET BUILDINGS,
à LONDRES.

La Societá denominata "WILLIAMS' PERRAN FOUNDRY CO." non ha giá fatto stampare questo suo Catalogo nell' intento di fornire informazioni dettagliate sul prezzo e sulla forza delle macchine ch'essa fabrica. Nell' industria mineralogica ed Idraulica, che mutano tanto nei dettagli, e che devono essere regolate da svariate circostanze locali d'ogni specie, sarebbe difficilissimo il compilare una tavola di prezzi.

Lo scopo della società é puramente quello di dare, agli Ingegneri meccanici ed altri, tanto all' estero che nell' interno, un' idea approssimativa del genere di macchine che fornisce, e che un' esperienza acquistata in tre quarti di secolo, l' ha messa in grado di portarle all' attuale alto grado di perfezione che le fa apprezzare in tutti i paesi nei quali esse vengono impiegate.

La Societá WILLIAMS' PERRAN FOUNDRY CO. sará ben felice di spedire dietro richiesta, in qualsiasi parte del Globo, estimi, piani, disegni e specifiche delle macchine che costruisce, non che di fornire i prezzi di qualsiasi macchina da costruirsi dietro disegni che le venissero inviati.

INDIRIZZARSI, A LONDRA,
No. 1 e 2 GREAT WINCHESTER STREET BUILDINGS.

La Sociedad intitulada "WILLIAMS' PERRAN FOUNDRY CO.," no han hecho imprimir este Catalogo con la idea de dar informes detallados sobre el costo y fuerza de la Maquinaria por ellos fabricada. En las industrias Mineras e hidraulicas que varian tanto en sus detalles y en las circumstancias locales, seria muy deficil establecer una tarifa de precios.

Su objeto es unicamente presentar à los Ingeníeros, mecanicos y otros, del Paiz como del Estrangero con quienes la Sociedad no haya tenido relaciones, una idea aproximativa de la clase de maqinas que la Sociedad ofrece al publico, y que una experiencia de cerca de setenta y cinco años la permita de producir en el mas alto grado de perfeccion para todos los Paises, en donde esta clase de Maquinas se emplean.

La Sociedad WILLIAMS' PERRAN FOUNDRY CO. tendrán siempre mucho gusto en remitir a cualquier parte del mundo, los mapaplanos, presupuestos, y especificaciones, qui se la piden, ò daran los presupuestos de los deseños que se le presenten.

1 y 2 GREAT WINCHESTER STREET BUILDINGS,
LONDRES.

PERRANARWORTHAL, CORNWALL,

AND

1 & 2 GREAT WINCHESTER STREET BUILDINGS, LONDON, E.C.

CONTENTS.

PERRANARWORTHAL, CORNWALL,
AND
1 & 2 GREAT WINCHESTER STREET BUILDINGS, LONDON, E.C.

MAKERS OF

BRITISH AND FOREIGN MINING MACHINERY,

ALL CLASSES OF

LAND ENGINES,

OF THE LARGEST SIZES.

Machinery for supplying water to Towns and for Irrigation, Drainage, and Sewage, according to the Designs and Specifications of Engineers.

PUMPING MACHINERY,

FOR EMPTYING

MINES, DOCKS, AND CANALS.

WATER WHEELS, MILL GEARING, HEAVY SHAFTING, CORNISH AND OTHER BOILERS. &c.

Below is a list of a few of the many Engines and Pumps, &c., made by the Perran Foundry Company :—

DESCRIPTION.	DIAMETERS OF CYLINDERS.		BUILT FOR.	DESTINATION.
Engine and Pump	64 inch	....	Vauxhall Water Works	London.
Ditto	144 ,,	....	Haarlem Lake Drainage	Holland.
Ditto	70 ,,	....	Grimsby Docks	Lincolnshire.
3 Engines and Pumps	80 ,,	.... ⎫	Burra Burra Mines	South Australia.
1 Engine and Pump	50 ,,	.... ⎬		
Ditto	85 ,,	....	Plymouth Iron Works	Wales.
Ditto	60 ,,	....	South Australia Copper Company	South Australia.
Pair of Engines and Pump	36 ,,	....	Water Works	Madrid.
Steam Blast Engines	72 ,,	.... ⎫	Ebbw Vale Steel and Iron Company Limited	Wales.
,, ,, ,,	144 ,,	.... ⎬		
Engine	85 ,,	....	Mold Mines	Wales.
Double Acting Engine and Pumps	85 ,,	....	New Bowson Colliery	Gloucestershire.
Engine and Pump	60 ,,	....	Stiperstone Mine	Shrewsbury.
Stationary Engines	16 ,,	....	Messrs. Bibbey & Company	Italy.
Double Acting Engine and Pump	36 ,,	....	Water Works	Lincoln.
Engine and Pumps	42 ,,	....	Torksey Drainage	Lincolnshire.
Pair Horizontal Engines	12 ,,	....	Maidstone Gas Works	Maidstone.

PERRANARWORTHAL, CORNWALL,

AND

1 & 2 GREAT WINCHESTER STREET BUILDINGS, LONDON, E.C.

WILLIAMS' PERRAN FOUNDRY CO.

ELEVATION OF PUMPING ENGINE.

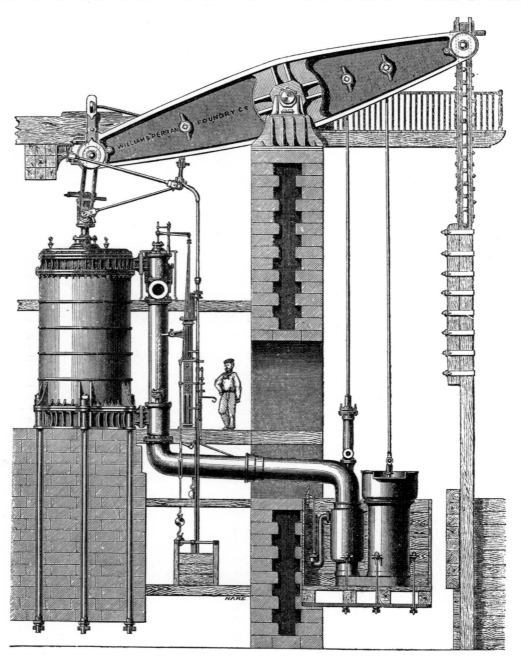

BUILT FOR ST. DAY UNITED MINES.

This Engine, made by us for the **St. Day United Mining Company** (now called the Poldice Mines), is a good example of the Cornish Engine as at present made, and gives a satisfactory idea of the general proportion of these **Engines.** It has been working under a load of 126,000 lbs. for the past eight years, and during the winter time has been kept continuously at work for six months, at an average rate of nearly eight strokes per minute, doing its work to the satisfaction of all concerned. **Estimates** on application.

August, 1870.

Williams' Perran Foundry Co.

PERRANARWORTHAL, CORNWALL,

AND

1 & 2 GREAT WINCHESTER STREET BUILDINGS, LONDON, E.C.

ELEVATION OF TAYLOR'S PUMPING ENGINE.

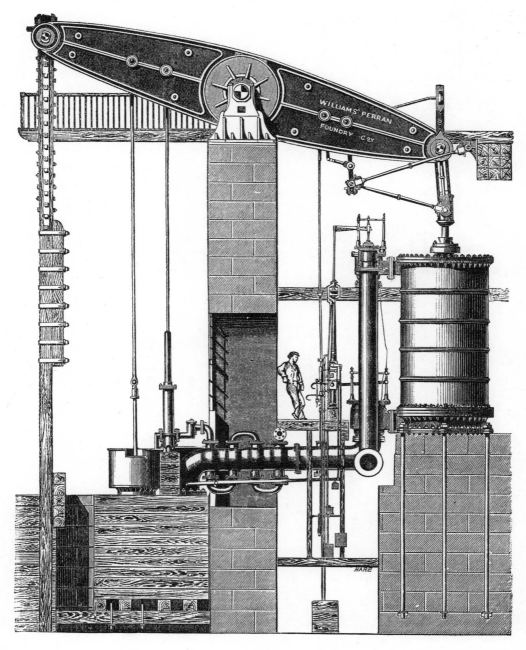

This is a representation of the famous Taylor's Engine, manufactured by us in 1840. It has been working for a period of 30 years, and notwithstanding this long service, made during a great part of last winter 9 (11 feet) strokes per minute. The average duty of this Engine during the years 1841 and 1842 was, according to Mr. Farey, 95,750,000 lbs. of water raised 1 foot high by the combustion of 1 bushel of coals, being a consumption of less than 2 lbs. of coal per horse-power per hour; and as stated on next page this Engine has been registered as high as 107,000,000 lbs. of water lifted 1 foot high with the same expenditure of fuel, being the highest duty on record. **Estimates** on application.

August, 1870.

Williams' Perran Foundry Co.

PERRANARWORTHAL, CORNWALL,
AND
1 & 2 GREAT WINCHESTER STREET BUILDINGS. LONDON, E.C.

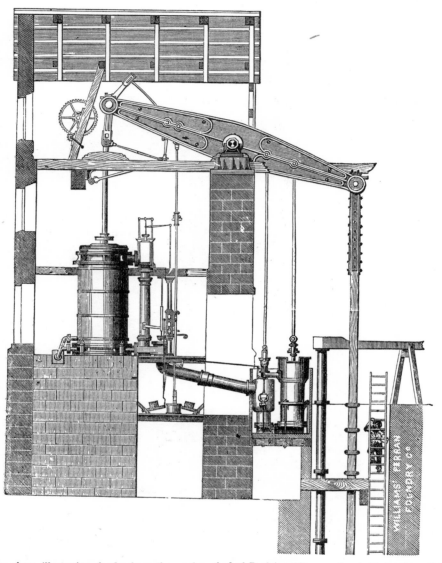
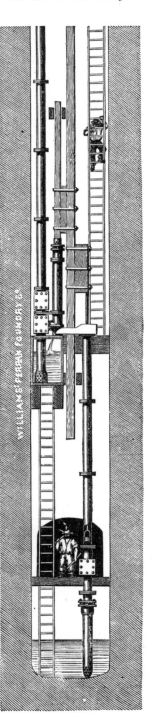

WHIM ENGINE,
WITH WINDING CAGE ATTACHED.

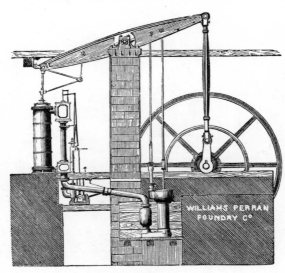

The accompanying illustration represents a side elevation of a Whim Engine with Winding Cage attached. This is the ordinary Engine employed for raising the productions of the Mine, and is sometimes adapted for the threefold purposes of Winding, Crushing, and Pumping. These Engines are made in the best modern style and of any power required. Estimates for simple Winding Engines, for Engines adapted to other purposes, with pump, gearing, straps, and boilers complete, furnished on application. These Engines are well finished and made in the most approved style and of any power required. For Winding, it is only necessary to state the height and daily tonnage of the lift. If for Crushing, state the number of tons and description of material required to be crushed in a given time; and if for Pumping, the depth of mine and quantity of water (in gallons) to be raised per hour.

SMALL WINDING ENGINE.

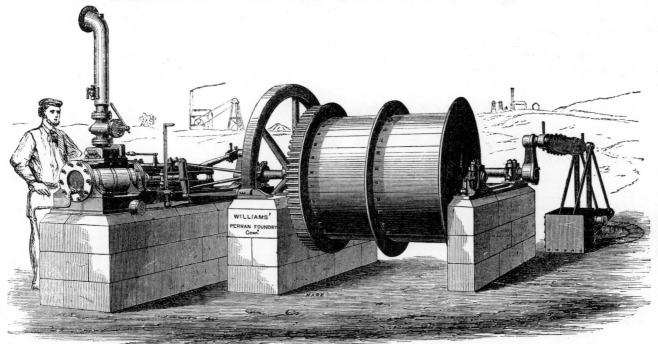

This illustration shows the general description of Engines sent abroad for the prosecution of small Mines. The Engine is of our usual Horizontal description with an extra sized Cornish Boiler, &c. The ore is raised by means of the double drum with a wire rope, so that one kibble is being raised whilst the other is lowered. The other shaft has a crank fixed on its end for working the pumps by means of the balance beam. These are so arranged that either the pumping or winding can be thrown out of gear to work either singly.

ESTIMATES ON APPLICATION.

IMPROVED HORIZONTAL HIGH-PRESSURE ENGINES.

These Engines are all made of the very best materials, and constructed after the most approved models, with a view to obtain the greatest results at the least expenditure. They are made sufficiently strong to yield more than double the nominal power if worked at a boiler pressure of, say 60 lbs. to 65 lbs.

It is needless to enter into a long descriptive specification of these Engines, as the high position they have held for so many years has established their reputation for great durability and economy in the consumption of fuel.

SPECIFICATION OF EACH SIZE GIVEN UPON APPLICATION.

Below are a few particulars of the smaller sizes, but **Williams' Perran Foundry Co.** construct Engines on this system up to the largest size made.

PARTICULARS AND PRICE OF ENGINES ONLY.

Nominal Horse Power.	Diameter of Cylinder.	Length of Stroke.	Price at Works.	If fitted with expansion valves extra			If with Feed Water Heater extra.			Cost of Packing.			Approximate weight with Fly Wheel.
	inches.	inches.	£ s. d.	£	s.	d.	£	s.	d	£	s.	d.	
4	6	10		9	0	0	9	0	0	2	0	0	
6	8	12		9	0	0	9	0	0	2	10	0	
8	9	14		9	0	0	9	0	0	3	0	0	
10	10	18		9	0	0	9	0	0	3	10	0	
12	12	20		14	0	0	14	0	0	4	0	0	
14	14	24					14	0	0	4	10	0	
16	15	24					18	0	0	5	0	0	
18	16	30		Included in price of Engine.			18	0	0	5	10	0	
20	18	30					18	0	0	6	0	0	
25	20	36					25	0	0	6	10	0	
30	22	42					25	0	0	7	10	0	

(For Boilers see next page.)

PERRANARWORTHAL, CORNWALL,
AND
1 & 2 GREAT WINCHESTER STREET BUILDINGS, LONDON, E.C.

CORNISH BOILERS.

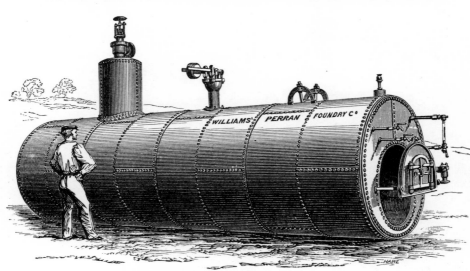

These Boilers are made of the best Staffordshire Iron, with Furnaces of Low Moor Iron, and are proved to any specified pressure by Hydraulic Power. They are constructed with the greatest care, and as **Williams' Perran Foundry Co.** have made during the last few years great numbers of Boilers of this class they feel confidence in stating that they can, both in price and quality, compete successfully with any other makers.

Williams' Perran Foundry Co. will furnish estimates for Boilers of any size, to stand any required pressure, according to the drawings and specifications of Engineers.

The Works being situated on the sea coast offer special facilities for shipment without heavy land carriage. Boilers are also constructed in pieces for facility of transport, and being marked and numbered are easily put together on arrival at their destination.

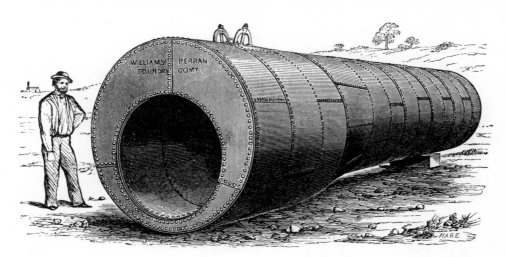

The prices given below are at the Works, and are subject to alteration according to the variation in the price of iron. **Williams' Perran Foundry Co.** also give estimates for these Boilers at per ton.

4 Horse-Power Boiler, without Fittings .. ..£
  ditto   ditto   with Fittings, inclusive of Safety-valves, one with Spring Balance, Blow off Cocks, Water Gauge, Steam Gauge, Mud Holes, Flue Doors, Man Holes, Furnace Doors and Frames, Damper and Pressure Gauge, Bearing and Furnace Bars, Weights, Pullies, Chains, &c., extra .. .. .. £

6 Horse-Power Boiler, without Fittings .. .. £			16 Horse-Power Boiler, without Fittings .. .. £		
	ditto	ditto	with Fittings, as above, extra .. £	ditto	ditto	with Fittings, as above, extra .. £
8	ditto	ditto	without Fittings .. .. £	18 ditto	ditto	without Fittings .. .. £
	ditto	ditto	with Fittings, as above, extra .. £	ditto	ditto	with Fittings, as above, extra .. £
10	ditto	ditto	without Fittings .. .. £	20 ditto	ditto	without Fittings .. .. £
	ditto	ditto	with Fittings, as above, extra .. £	ditto	ditto	with Fittings, as above, extra .. £
12	ditto	ditto	without Fittings .. .. £	25 ditto	ditto	without Fittings .. .. £
	ditto	ditto	with Fittings, as above, extra .. £	ditto	ditto	with Fittings, as above, extra .. £
14	ditto	ditto	without Fittings .. .. £	30 ditto	ditto	without Fittings .. .. £
	ditto	ditto	with Fittings, as above, extra .. £	ditto	ditto	with Fittings, as above, extra .. £

PERRANARWORTHAL, CORNWALL,
AND
1 & 2 GREAT WINCHESTER STREET BUILDINGS, LONDON, E.C.

WATER WHEEL DRIVING STAMPS,

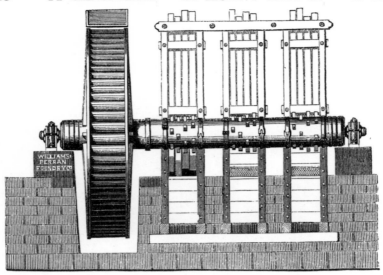

The above Sketch represents a Water Wheel working a Stamps, consisting of 12 Heads, for breaking up Ore and preparing it for dressing, or separating the Mineral from the Rubbish. In some of the large Cornish Mines, Steam Power is used to work a great number of these Heads or Stampers, which are usually constructed with long iron lifters instead of wood. It will be seen from the illustrations that Lifters are fixed in the Barrel, which is caused to revolve by means of a Water Wheel or other motive power; there are also Lifters in the uprights, which are secured to the Stampers; the Heads are thus raised from 9 to 12 in. high, consecutively, and falling their full weight (from 3 to 5 cwt. each) on the Ore, break it in pieces and reduce it to powder.

This illustration shows another arrangement of the Battery, each set consisting of 6 Heads of Stamps.

Though from time to time a great many machines for Crushing and Grinding in different ways have been devised, with a view to supersede the Cornish Stamps, nothing has, however, yet been found which does the work either for Gold, Silver, or Tin, in such a cheap and effectual manner.

PRICES ON APPLICATION.

PERRANARWORTHAL, CORNWALL,

AND

1 & 2 GREAT WINCHESTER STREET BUILDINGS, LONDON, E.C.

SHORT SHANK STAMP HEADS,
TO FIT WOOD LIFTERS,

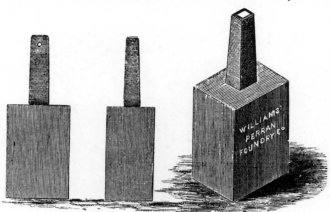

The **Williams' Perran Foundry Co.** have Patterns of the various sizes of Stamp Heads used in the British and Foreign Mines, and can supply large quantities on the shortest notice, made of the best White Iron, chilled or steel as required.

WATER WHEELS.

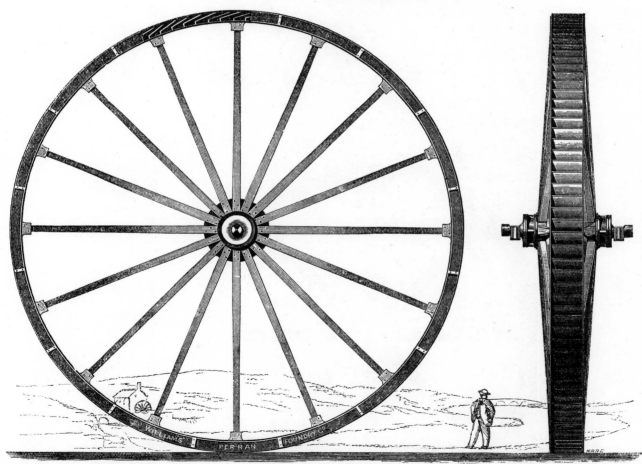

Williams' Perran Foundry Co. make Water Wheels of every description, both in wood and iron. The simple natural power of running or falling water is capable of application by means of the Water Wheel to an almost unlimited variety of purposes, for the mine, the supply of water to towns, villages, and farms, for irrigating wide tracts of land, for driving mills, supplying power to factories, cranes, cutting timber, crushing ores, lifting and winding heavy weights, either directly or with gearing attached; in fact, there is no limit, according to its power, to the application of this most economical of all agents wherever it is placed.

PERRANARWORTHAL, CORNWALL,
AND
1 & 2 GREAT WINCHESTER STREET BUILDINGS, LONDON, E.C.

WATER WHEELS—*continued.*

Williams' Perran Foundry Co. will, on receipt of the necessary information (such as the fall of water, the depth, width, and rate of stream to be utilized, or where it is only necessary to employ a portion of the force of the stream, the amount of work required to be done by the wheel, also for what class of wheel the position is adapted, and such other data as will suggest themselves to the Engineer) furnish estimates, not only for the wheel and gearing, but for all such portions of the machinery as may come within the Company's province as Engineers and Ironfounders.

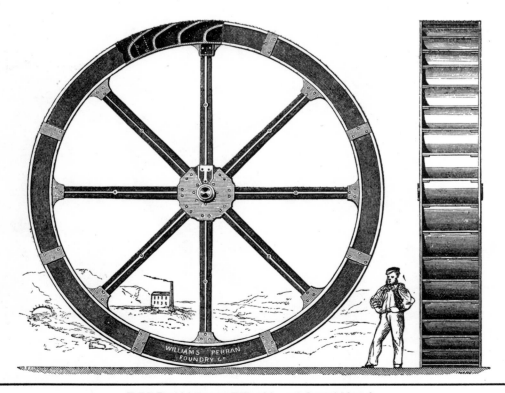

CAPSTAN,

For raising or lowering heavy weights in Shafts, such as Pumps, H. and Door Pieces, parts of Main Connecting Rods, &c. Cast Iron Axles, Centre Pieces for Arms, Heads, and Foot Blocks, Bolts, Plates, and Straps

may be had ready to mount the wood-work, or the whole may be fitted together at the Works, marked and taken to pieces for transport.

Prices for different sizes on application.

STRONG SINGLE PURCHASE CRABS,
To lift from 2 to 5 Tons, with or without breaks, made in the best manner for general use.

POWERFUL DOUBLE PURCHASE CRABS.
To lift from 4 to 12 Tons, with ratchet wheels, shifting dogs, pinion shafts to slide, so as to work either single or double purchase; these are built in a most substantial manner, and thoroughly adapted for heavy work.

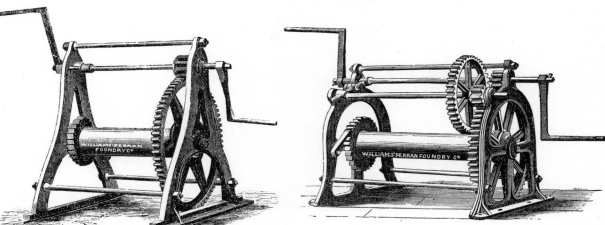

WINCHES OF ANY POWER.—PRICES ACCORDING TO WEIGHTS TO BE LIFTED.

WILLIAMS' PERRAN FOUNDRY CO.

IMPROVED PORTABLE CRANE.

This Crane is counter-balanced and works on the centre, and is extremely useful for loading and unloading at Railway Stations, Breweries, Large Factories and Public Works.

Prices.	£ s. d.
1 ton	..
2 ,,	..
3 ,,	..
4 ,,	..
5 ,,	..
10 ,,	..

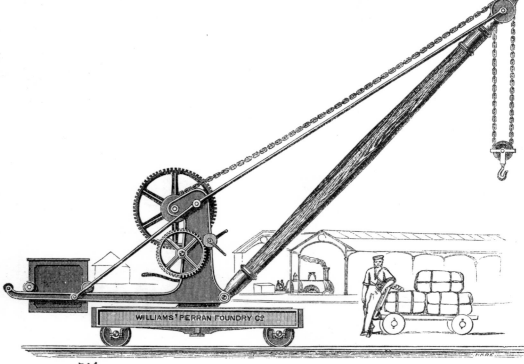

WROUGHT IRON CRANE.

These Jib Cranes for warehouses are fitted with Top Carriage Step Bracket, and Chain Pulley complete, all ready to fix.

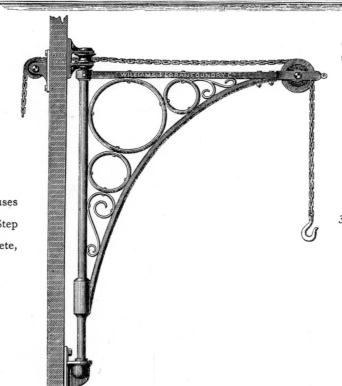

Prices.			£	s.	d.
1 ton	..	..	..	..	
30 cwt.	..	..	..	..	
2 tons.	..	..	..	..	
3 ,,	..	..	..	..	
4 ,,	..	..	..	..	

PERRANARWORTHAL, CORNWALL,

AND

1 & 2 GREAT WINCHESTER STREET BUILDINGS, LONDON, E.C.

LIFTING CRANE,

FOR WHARVES, PIERS, GOODS STATIONS AND DOCKS, &c.,

Williams' Perran Foundry Co. also construct Derrick Cranes of any size or power, and build Cranes to any special pattern required by Engineers and Contractors; and will, on receipt of drawing and specification, furnish an estimate for any number.

Prices.	£	s.	d.
1 ton ..			
30 cwt. ..			
2 tons ..			
3 ,, ..			
5 ,, ..			
8 ,, ..			
10 ,, ..			
13 ,, ..			
15 ,, ..			
20 ,, ..			

CAST AND WROUGHT TANKS,

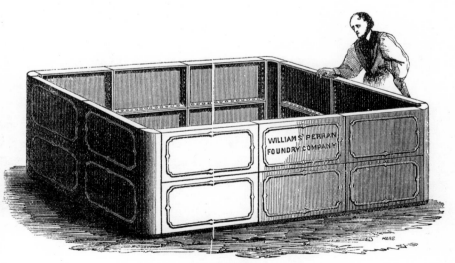

IRON TANK.

For storing or holding water for Railway Stations, Gaols, Barracks, Factories and all Isolated Buildings.

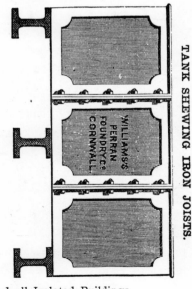

TANK SHEWING IRON JOISTS.

PERRANARWORTHAL, CORNWALL,

AND

1 & 2 GREAT WINCHESTER STREET BUILDINGS, LONDON, E.C.

PORTABLE ENGINES AND CENTRIFUGAL PUMPS.

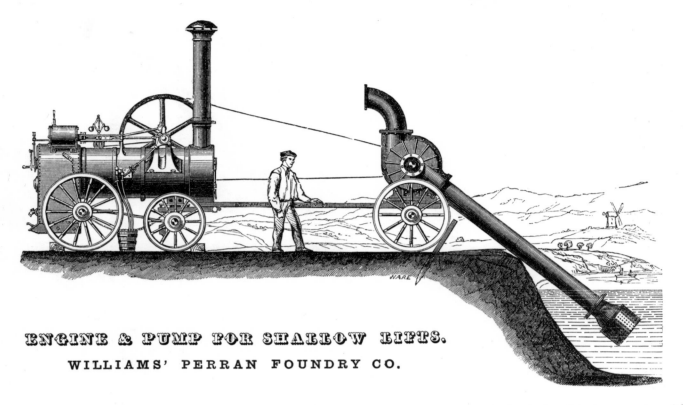

ENGINE & PUMP FOR SHALLOW LIFTS.

WILLIAMS' PERRAN FOUNDRY CO.

There are many circumstances under which the employment of a Portable Pump and Engine is found to be a most useful adjunct to Mining operations, either for assisting to divert the course of a stream, the debris of which may contain valuable ores, or for supplying water from the bed of a stream for the Crushers, and for other purposes. The Engine, when not used for pumping, can be employed for cutting timber, shores, planks, &c., or for driving, where practicable, the stamps, or for many other operations that will suggest themselves to the practical Miner. The Engines and Pumps supplied by **Williams' Perran Foundry Co.** embody the most improved patents, and are fitted in the best possible manner. The Engines are specially suited for foreign work. Where desired, the fire-boxes of Engines are made extra large for burning wood, dung, &c., &c., at a small increased cost.

The above illustration is a good arrangement; the Pump is fitted on a carriage with two wheels, shafts or pole. These shafts rest on the frame of the Portable Engine, and keep the Pump firm and the driving-strap straight. The driving-strap runs direct from fly-wheel of Engine to the Pump-pulley. In the Price List given (No. 1) on next page, the diameter of pulley and number of pump revolutions are given. This will show the speed at which Engines must be driven, the speed of course varying according to the diameter of fly-wheel of Engine. But where a purchaser has an Engine, and a Pump only is wanted, it is better always to state the diameter of fly-wheel of Engine and the name of maker of such Engine.

The Suction-pipe is attached diagonally to the Pump so as to enable it to be readily placed in the water. The foot valve (with a grating underneath to prevent the entrance of weeds, leaves, stones, pieces of timber, &c., into the Pump) is placed at the bottom of the pipes. The pipes are of wrought iron, and can be laid on the carriage when required to be moved from place to place. The Engines, when it is desired, can at a small increased cost be fitted with Link Motion Reversing Gear, and with Powerful Brakes to hind wheels for hilly countries. Covers or awnings can be also fitted to shade the driver in hot climates.

PERRANARWORTHAL, CORNWALL,

AND

1 & 2 GREAT WINCHESTER STREET BUILDINGS, LONDON, E.C.

PRICES OF ENGINES AND PUMPS ON TWO-WHEELED CARRIAGE.—No. 1.

10 FEET LIFT.

Horse Power of Engine ..	6	8	10	12	14	20	25	30
Letter or No. of Pump ..	B^2	B^3	B^4	B^4	B^4	B^5	B^6	
Diameter of Pump pulley	10″	12″	13″	13″	13″	16″	20″	
Number of Pump revolutions per minute	394	364	328			275	226	
Gallons raised per minute	1000	1450	1900	2400	2550	4200	5500	
	£ s. d.	£ s. d.	£ s. d.	£ s. d.	£ s. d.	£ s. d.	£ s. d.	£ s. d.
Price of Engine	180 0 0	210 0 0	240 0 0	300 0 0	335 0 0	445 0 0		
Packing Engine for Export	6 0 0	8 0 0	10 0 0	12 0 0	14 0 0	20 0 0		
Price of Pump with wrought iron pipes complete	73 0 0	80 0 0	98 0 0	102 0 0	112 0 0	133 0 0	200 0 0	
Packing Pump for Export	1 16 0	2 3 0	2 12 0	2 18 0	3 3 0	3 15 0	5 2 0	
Price of Leather Driving-band, 60 feet long	15 0 0	18 10 0	20 15 0	22 0 0	23 10 0	40 10 0	49 0 0	
Total cost complete								

20 FEET LIFT.

Horse Power of Engine ..	6	8	10	12	14	20	25	30
Letter or No. of Pump ..	No. 4	B^1	B^2	B^3	B^3	B^4	B^5	
Diameter of Pump pulley	10″	9″	10″	12″	12″	13″	16″	
Number of Pump revolutions per minute	495	655	570	510		435	390	
Gallons raised per minute	530	700	1000	1400	1500	2200	3000	
	£ s. d.	£ s. d.	£ s. d.	£ s. d.	£ s. d.	£ s. d.	£ s. d.	£ s. d.
Price of Engine	180 0 0	210 0 0	240 0 0	300 0 0	335 0 0	445 0 0		
Packing Engine for Export	6 0 0	8 0 0	10 0 0	12 0 0	14 0 0	20 0 0		
Price of Pump with wrought iron pipes complete	61 0 0	73 0 0	81 0 0	90 0 0	105 0 0	112 0 0	148 0 0	
Packing Pump for Export	1 6 0	1 12 0	1 16 0	2 3 0	2 3 0	2 12 0	3 15 0	
Price of Leather Driving-band, 60 feet long	15 0 0	12 10 0	15 0 0	18 10 0	19 10 0	20 15 0	40 10 0	
Total cost complete								

30 FEET LIFT.

Horse Power of Engine ..	6	8	10	12	14	20	25	30
Letter or No. of Pump ..	No. 3	No. 4	B^1	B^1	B^2	B^3	B^4	
Diameter of Pump pulley	6″	10″	9″	9″	10″	12″	13″	
Number of Pump revolutions per minute	835	608	795		695	630	536	
Gallons raised per minute	300	430	600	750	850	1100 to 1400	1500 to 1900	
	£ s. d.	£ s. d.	£ s. d.	£ s. d.	£ s. d.	£ s. d.	£ s. d.	£ s. d.
Price of Engine	180 0 0	210 0 0	240 0 0	300 0 0	335 0 0	445 0 0		
Packing Engine for Export	6 0 0	8 0 0	10 0 0	12 0 0	14 0 0	20 0 0		
Price of Pump with wrought iron pipes complete	48 0 0	64 0 0	79 0 0	79 0 0	89 0 0	99 0 0	122 0 0	
Packing Pump for Export	1 0 0	1 6 0	1 12 0	1 15 0	1 16 0	2 3 0	2 12 0	
Price of Leather Driving-band, 60 feet long	8 15 0	15 0 0	12 10 0	13 15 0	15 0 0	18 10 0	20 15 0	
Total cost complete								

In ordering Engine and Pump always give the Letter and No. of Pump required.

THE ABOVE PRICES ARE DELIVERED IN LONDON. SPECIAL PRICES FOR THE LARGER ENGINES AND PUMPS.

CENTRIFUGAL PUMPS ON BED PLATES ONLY,

FOR LIFTING LARGE VOLUMES OF WATER FROM 10 TO 30 FEET.

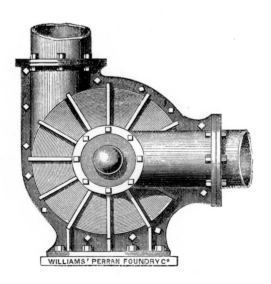

The prices and particulars given on page 21 are for the same Pumps as described on page 18, but on bed plate only, ready for bolting down in any required position, to be driven by Mining Engine, Water Wheel, or other power, as the case may be.

The total cost is not filled in as the number of feet of leather driving band required must depend upon the distance between driving wheel of power and the pump pulley. By allowing for double this distance, and the diameter of drums, the cost of band can be calculated, as the price is given at per foot. By adding this to the other figures the total cost is obtained.

PERRANARWORTHAL, CORNWALL,

AND

1 & 2 GREAT WINCHESTER STREET BUILDINGS, LONDON, E.C.

PRICES OF PUMPS, &c., WITHOUT TWO-WHEELED CARRIAGE AND ENGINE.
No. 2.

10 FEET LIFT.

Horse Power required ...	6	8	10	12	14	20	25	30
Letter or No. of Pump ...	B²	B³	B⁴	B⁴	B⁴	B⁵	B⁶	
Diameter of Pump pulley...	10″	12″	13″	13″	13″	16″	20″	
Number of Pump revolutions per minute... ...	394	364	328			275	226	
Gallons raised per minute	1000	1450	1900	2400	2550	4200	5500	
	£ s. d.	£ s. d.	£ s. d.	£ s. d.	£ s. d.	£ s. d.	£ s. d.	£ s. d.
Price of Pump with wrought iron pipes ...	56 10 0	62 0 0	77 0 0	87 0 0	97 0 0	105 0 0	165 0 0	
Packing Pump for Export	1 16 0	2 3 0	2 12 0	2 12 0	2 12 0	3 15 0	5 2 0	
Price of Leather Driving-band ... ... per foot	0 5 0	0 6 2	0 6 11	0 6 11	0 7 8	0 13 6	0 16 4	
Total cost complete								

20 FEET LIFT.

Horse Power required ...	6	8	10	12	14	20	25	30
Letter or No. of Pump ...	No. 4	B¹	B²	B³	B³	B⁴	B⁵	
Diameter of Pump pulley..	10″	9″	10″	12″	12″	13″	16″	
Number of Pump revolutions per minute... ...	495	655	570	510		435	390	
Gallons raised per minute	530	700	1000	1400	1500	2200	3000	
	£ s. d.	£ s. d.	£ s. d.	£ s. d.	£ s. d.	£ s. d.	£ s. d.	£ s. d.
Price of Pump with wrought iron pipes ...	50 0 0	59 0 0	64 10 0	72 0 0	87 0 0	91 0 0	120 0 0	
Packing Pump for Export	1 6 0	1 12 0	1 16 0	2 3 0	2 3 0	2 12 0	3 15 0	
Price of Leather Driving-band ... ... per foot	0 5 0	0 4 2	0 5 0	0 6 2	0 6 6	0 6 11	0 13 6	
Total cost complete								

30 FEET LIFT.

Horse Power required ...	6	8	10	12	14	20	25	30
Letter or No. of Pump ...	No. 3	No. 4	B¹	B¹	B	B³	B⁴	
Diameter of Pump pulley...	6″	10″	9″	9″	10″	12″	13″	
Number of Pump revolutions per minute... ...	835	608	795		695	630	536	
Gallons raised per minute	300	430	600	750	850	1100 to 1400	1500 to 1900	
	£ s. d.	£ s. d.	£ s. d.	£ s. d.	£ s. d.	£ s. d.	£ s. d.	£ s. d.
Price of Pump with wrought iron pipes ...	39 0 0	53 0 0	65 0 0	68 0 0	72 10 0	81 0 0	101 0 0	
Packing Pump for Export	1 0 0	1 6 0	1 12 0	1 15 0	1 16 0	2 3 0	2 12 0	
Price of Leather Driving-band ... ... per foot	0 2 11	0 5 0	0 4 2	0 4 2	0 5 0	0 6 2	0 6 11	
Total cost complete								

In ordering Pump always give the Letter and No. of Pump required.

THE ABOVE PRICES ARE DELIVERED IN LONDON.

PERRANARWORTHAL, CORNWALL,

AND

1 & 2 GREAT WINCHESTER STREET BUILDINGS, LONDON, E.C.

IMPROVED PORTABLE ENGINES.

SINGLE-CYLINDER ENGINES, from 6 to 10 Horse-Power.
DOUBLE-CYLINDER ENGINES, from 10 to 20 Horse-Power.

These Engines are especially constructed to suit the wants of the Australian Colonies, the Foreign British Possessions, and all those Countries where such power is wanted, and where the greatest results must be attained by the smallest outlay both of fuel and manual labor. It is useless to give a long written description of these Engines, now employed in almost every country, either for Agricultural and Mining operations, or as auxiliaries in the construction of Public Works, but a specification of each size can always be had on application. It is sufficient to say these Engines are built on the latest and most improved principles, and of the very best material. They have plenty of boiler power, and will make steam (with a like amount of fuel) faster than any other Engine of this class. They are at once the most durable and economical of portable powers.

PRICES AND PARTICULARS OF ENGINES.

HORSE-POWER.	S. 5	S. 6	S. 8	S. 10	D. 10	D. 12	D. 14	S. 16	D. 20
Diameter of Cylinder	7¼"	8½"	9½"	11"	7¼"	7¾"	8½"	13¼"	11"
Length of Stroke	10"	13"	13"	13"	13"	13"	13"	16"	15"
Diameter and face of Fly-Wheel	4' 6" × 5"	5' 0" × 5½"	5' 0" × 7"	5' 0" × 7"	5' 0" × 7"	5' 0" × 7"	5' 0" × 7"	6' 0" × 7"	5' 0" × 9"
Number of Revolutions	160	120	120	120	120	120	120	96	100
Coal per hour	35 lbs.	42 lbs.	56 lbs.	70 lbs.	70 lbs.	84 lbs.	98 lbs.	112 lbs.	140 lbs.
Water per hour	6½ gals.	7¾ gals.	10¼ gals.	12 gals.	12 gals.	15 gals.	18 gals.	20½ gals.	25 gals.
Tested Pressure	200	200	200	200	200	200	200	200	200
Working Pressure	80	80	80	80	80	80	80	80	80
Net Weight of Engine	2 9 0 0	3 4 0 0	3 13 0 0	4 0 0 0	4 5 0 0	4 15 0 0	5 5 0 0	6 0 0 0	7 15 0 0
Gross Weight with Case	3 0 0 0	3 15 0 0	4 5 0 0	4 15 0 0	5 0 0 0	5 5 0 0	6 0 0 0	7 0 0 0	8 10 0 0
Cubic Contents Packed	220	356	413	430	430	460	490	520	540
Price in London	£165	£180	£210	£240	£260	£300	£335	£335	£445
Extra for Large Fire Box (for burning wood)	£2 10	£3	£4	£5	£5	£6	£7	£8	£10
Charge for Packing for Export	£5	£6	£8	£10	£10	£12	£14	£16	£20

PERRANARWORTHAL, CORNWALL,
AND
1 & 2 GREAT WINCHESTER STREET BUILDINGS, LONDON, E.C.

PORTABLE ENGINE AND WINDING GEAR FOR MINES.

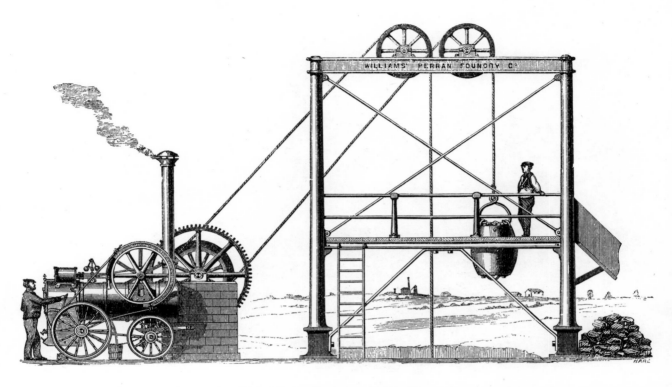

The above shows another application of the Portable Engines, which for light weights and lifts not too deep is often of great service. The Engine is fitted with all necessary winding gear, brake and reversing gear. It is only necessary for purchasers wanting the apparatus, to state the depth of lift and the quantity required to be raised per hour, when all particulars can be at once given. We give prices and particulars of the most usual sizes below, exclusive of iron work, columns, girders, platform, &c., ropes and kibbles, but including working drawings for erection.

Horse-power.	S. 8	S. 10	D. 10	D. 12	D. 14	S. 16	D. 20
Diameter of Cylinder	9½″	11″	7¼″	7¾″	8½″	13¼″	11″
Length of Stroke	13″	13″	13″	13″	13″	16″	15″
Diameter and Face of Fly-Wheel	5′ 0″×7″	5′ 0″×7″	5′ 0″×7″	5′ 0″×7″	5′ 0″×7″	6′ 0″×7″	5′ 0″×9″
Number of Revolutions	120	120	120	120	120	96	100
Coals per hour	56 lbs.	70 lbs.	70 lbs.	84 lbs.	98 lbs.	112 lbs.	140 lbs.
Water per hour	10¼ gals.	12 gals.	12 gals.	15 gals.	18 gals.	20½ gals.	25 gals.
Tested Pressure	200	200	200	200	200	200	200
Working Pressure	80	80	80	80	80	80	80
Net Weight of Engine	3 13 0 0	4 0 0 0	4 5 0 0	4 15 0 0	5 5 0 0	6 0 0 0	7 15 0 0
Gross Weight with Case	4 5 0 0	4 15 0 0	5 0 0 0	5 5 0 0	6 0 0 0	7 0 0 0	8 10 0 0
Cubic Contents Packed	413	430	430	460	490	520	540
Price in London	£210	£240	£260	£300	£335	£335	£445
Extra for Large Fire Box for Burning Wood	£4	£5	£5	£6	£7	£8	£10
Charge for Packing for Export	..	..	..	..	..	..	..
Price with Reversing Gear and Brake to Fly-Wheel, and one set of Winding Gear including Crotch, Connecting Rod, 2 Over-head Pulleys with bearings, 2 Spring Hooks, 2 5-feet Lengths of ½″ Short Link best Crane Chain, with eyes to attach to Steel Wire Rope (exclusive of Iron-work Ropes and Kibbles.)	£375	£405	£460	£500	£540	£545	£665
Charge for Packing Engine and Gear for Export	£15	£15	£15	£17	£19	£21	£25

PERRANARWORTHAL, CORNWALL,
AND
1 & 2 GREAT WINCHESTER STREET BUILDINGS, LONDON, E.C.

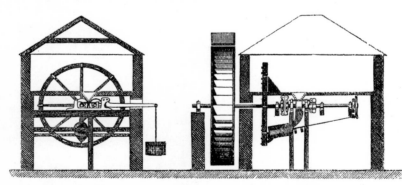

REVOLVING STAMPS.

BATTERY OF 4 HEADS.   **BATTERY OF 8 HEADS.**

These can be driven either by stationary or portable Engines, or by horse or bullock power. The whole is cast in small pieces for facility of transport over mountain roads. These are the most effectual stamps made. They are supplied for Australia, New Zealand, Chili, Mexico, Brazil, California, Nova Scotia, Tasmania, and the Colorado territory.

PRICES ON APPLICATION.

WILLIAMS' PERRAN FOUNDRY CO.

STEAM CAPSTANS,

TO LIFT FROM TEN TO FIFTY TONS.

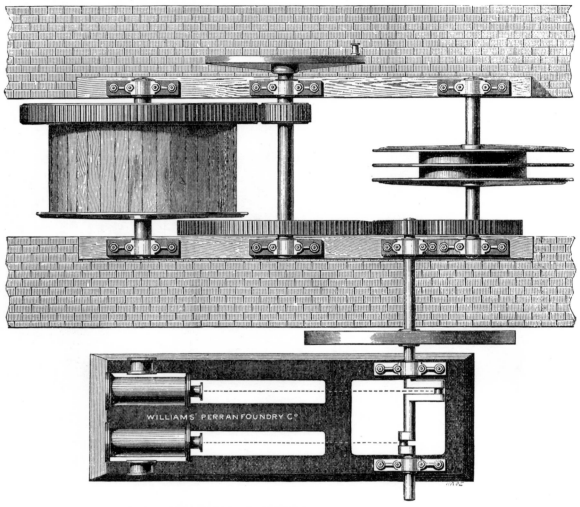

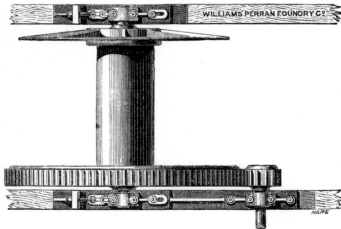

The Steam Capstan, shown in the two accompanying Cuts, which is simply a winch with a large barrel worked either by a pair of Engines and Gearing or by a single pinion and large wheel from the more powerful winding Engine, is now generally used for raising and lowering the Pump Rods, &c., &c. in the shaft, instead of the Capstan worked by manual labor, as shewn on page 15. As the large pumps, &c. take a considerable time to fix, it is very expensive to have a number of men employed night and day working the hand labor Capstan; and consequently the work is done more quickly and at less cost by the Capstans above shewn.

Prices, which depend on weights, &c., to be lifted, can be had on application.

FERRANARWORTHAL, CORNWALL,

AND

1 & 2 GREAT WINCHESTER STREET BUILDINGS, LONDON, E.C.

IMPROVED LOAM AND MORTAR MILL.

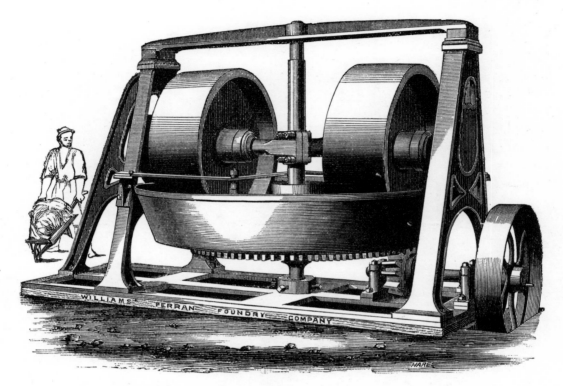

CAN BE WORKED BY ANY DESCRIPTION OF POWER WITH LEATHER BELT.

PARTICULARS OF MORTAR MILLS.

Horse-Power Required.	Diameter of Pan at top.	Size of Rollers.	Weight of each Roller.	Size of Pulley.	Speed of Pulley.	Speed of Pan.	Total Weight of Mill about.		Price in London.
			cwt.				Tons	cwts.	£
5	5' 0"	2' 8" × 10"	15	2' 0"	155	30	3	0	55
6	6' 0"	2' 7½" × 13½"	13	2' 0"	130	26	4	0	60
7	7' 0"	3' 6" × 15"	18	3' 0"	121	22	4	15	70
8	7' 6"	3 6" × 15"	22	3' 6"	110	20	5	15	75
10	9' 0"	3' 6" 18"	30	3' 6"	100	18	7	15	90

THE 5 FEET MILL IS ON TRAVELLING WHEEL.

PERRANARWORTHAL, CORNWALL,
AND
1 & 2 GREAT WINCHESTER STREET BUILDINGS, LONDON, E.C.

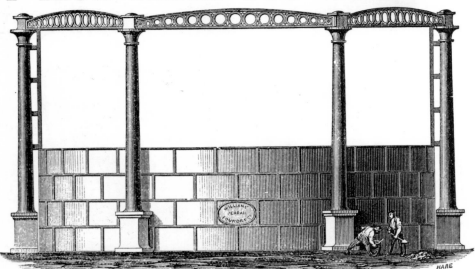

WEARING PARTS OF ENGINES, &c.

PLUNGER POLE, STUFFING BOX and GLAND, with BOLTS COMPLETE, as used in the CORNISH MINES.

These parts are required to be renewed oftener than some portions of the work, because, being working parts they are more subjected to wear. In

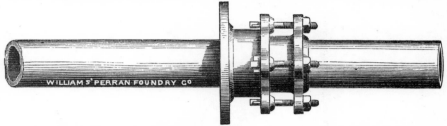

ordering these, the dimensions of the top flange of the Pole case should be given.

NOTE.—These are parts of the Plunger Lift.

WROUGHT IRON RISING MAIN PIPES, FOR PUMPS.

These Pipes are placed between the head of the Pump and the working barrels, to permit the Pump to be worked to a greater depth than 28-ft. below the surface or point of delivery; they are usually made in 10-ft. lengths, and are only about one-fifth of the weight of cast iron Pipes.

Diameter.				£	s.	d.
8½-in. to clear	8-in	Buckets	..	..	..	
9½-in.	„	9-in.	„	..	..	..
10½-in.	„	10-in.	„	..	..	..
13-in.	„	12-in.	„	..	..	..

Diameter.				£	s.	d.
16-in. to clear	15-in.	Buckets	..	..	..	
19-in.	„	18-in.	„	..	..	..
26.in.	„	24-in.	„	..	..	..

WORKING BARREL, CLACK DOOR PIECE, DOOR,
AND
WIND BORE.

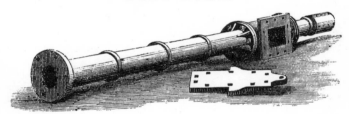

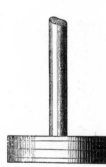

METALLIC PISTONS,

With Single or Double Rings, with or without Piston Rods. Piston Rods of best Faggotted Iron or Steel, turned and fitted by Machinery of the best modern construction.

BEAMS OR BOBS, MADE OF CAST OR WROUGHT IRON.

Plain or Ornamental, Close or Open. Plummer Blocks, Cast or Wrought Iron Gudgeons, Stools and Entablature Plates.

WEARING PARTS OF ENGINES, &c.—*continued.*
V BOBS, FOR FLAT RODS AND UNDERLAY SHAFTS.

These V Bobs are used in places where it is necessary to change the direction of the main rod or flat rods, so as to work the Pumps when the Shaft is perpendicular to a certain depth, and then sunk on the course of the Lode.

WROUGHT IRON MAIN CAPS, with SADDLE, BRASS, and BOLTS.

This is the connection from the outer end of the Engine Beam to the top of the Main Connecting Rod.   See end of Beam of Cornish Engine, pages 6, 7 & 8.

STAPLE AND GLAND.

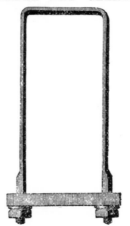

These are used to attach the different Plunger Poles to the Main Connecting Rod—from 4 to 9 or 10 are used—in proportion to the size of the Plunger Pole. See Mine Pumps.

REGULATING FEED NOZZLE.

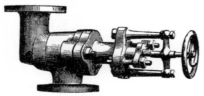

For Steam Boilers. With Brass Valve and Seat, Cross-head, and Adjusting Screw, complete.

IRON WORK FOR SCREW STOCKS.

These are fixed to wood posts, used in Mines for screwing Bolts, Nuts, Screws, Staples, &c.

CLACK SEAT AND VALVE.

These are the kind of Valves and Seats used in most of the Mine Pumps of Cornwall, the seatings of the H. or Door Piece is bored out to receive them,— they are made either of Brass or Iron.

SMALL FEED OR CHECK NOZZLE.

WEARING PARTS OF ENGINES, &c.—*continued.*

H PIECE AND DOOR PIECE.

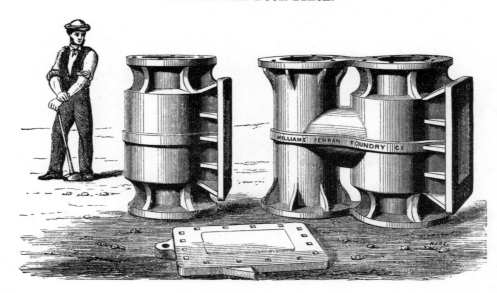

WING GUDGEONS.

To form the bearings of Water or other Wheels, of which the Axle is Wood. These Wings are let into the Wood, and confined with Wrought Iron Bands put on hot, which become tight by shrinking.

CRUSHER ROLLS,

Made of very hard metal, to crush Copper, Tin, Lead, and other Ores, as used in the British and Foreign Mines. See description of Crushing Mill, Page 24.

SEGMENT OF TOOTH WHEEL,

Made in Pieces for Foreign Countries where they have to be carried by Mules, and only a limited weight can be taken. Tooth and other large Wheels are made in segments when required.

BEVIL MORTICE WHEEL,

For Gearing with Wood Teeth.

TOOTHED WHEEL.

A large stock of Patterns of various sizes of Spur, Bevil, Mortice, and Segment Wheels.

WEARING PARTS OF ENGINES, &c.—*continued*.

SLUICE VALVES FOR GAS, WATER, AND STEAM.

DOUBLE SAFETY VALVE PIECE.

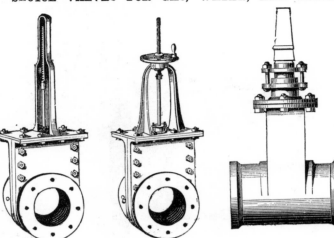

Made of all sizes, from 2-in. upwards.

With Lever and Weights, for Steam Boilers.

TWO VARIETIES OF DOUBLE BEAT VALVES AND SEATS.

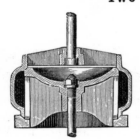

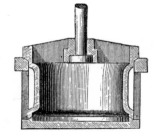

Boiler Nozzle, Steam Box, or Throttle Valve,

SECTION OF FEED PUMP,

A DOUBLE FEED NOZZLE.

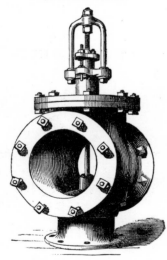

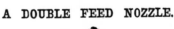

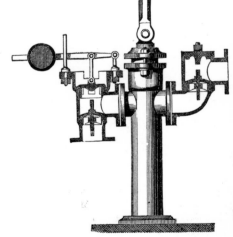

For stopping off the Steam from any Boiler when out of use.

With ground-in Spindle Valves and Seats, provided with a Safety Valve, so that no injury may arise by shutting off the Feed whilst the Pump is at work.

PERRANARWORTHAL, CORNWALL,
AND
1 & 2 GREAT WINCHESTER STREET BUILDINGS, LONDON, E.C.

WEARING PARTS OF ENGINES, &c.,—*continued.*

BRIGHT OR BLACK MILL SHAFTING, COUPLING, CLUTCHES, STRAP PULLEYS, PLUMMER BLOCKS, &c.,

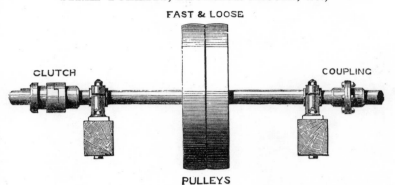

For all Mining and Manufacturing purposes.

LARGE PLUMMER BLOCKS AND BRASSES.

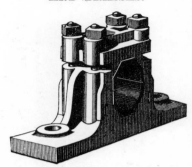

For Gudgeons, Water Wheel and Mill Shafts, fitted with Bolts, and bored to any required dimensions.

WROUGHT OR CAST IRON CRANKS AND SHAFTS,

For Land or Marine Engines, Heavy Forgings, by Powerful Steam and other Hammers, either Black or turned and fitted up Bright.

STRAIGHT FLY WHEEL AND OTHER SHAFTS,

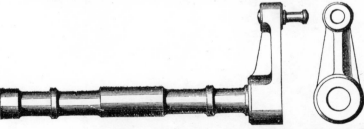

Of Cast or Faggotted Wrought Iron, Bright or Black, with or without Crank, Arms, or Drivers, Pins, Strap-heads, Brasses, &c. Also Fly Wheels either made whole or in Segments, with detached centres and arms.

BOLT ENDS, with V or Flat Thread Screws,

For holding down Bolts for Cylinders, &c., ready to shut up to any length required, also box or open-end spanners, to fit any size nuts or burrs.

Bolts and nuts of all sizes for Mines, Railways, Gas and Water Works, &c.

MINING TRAM WHEEL,

OF ANY PATTERN.

SADDLES OR CHAIRS,

Of various patterns, according to the section of the Rails.

PERRANARWORTHAL, CORNWALL,
AND
1 & 2 GREAT WINCHESTER STREET BUILDINGS, LONDON, E.C.

WEARING PARTS OF ENGINES, &c.—*continued*.

KIBBLES OF EVERY DESCRIPTION,

For bringing up the Ore and Rubbish or Attle from underground. Kibble Moulds of various sizes may be had for making the Plates into shape and manufacturing the Kibbles at the Mines.

LARGE PULLEY OR SHEAVE,

For top of Capstan, Shears, or Poppet Heads, for large Rope or Chain, as used in Mines for raising or lowering heavy weights in Shafts, &c.

LARGE FLAT ROPE PULLEY OR SHEAVE,

As used for the purposes of raising the produce of the Mine.

CARRYING PULLEY FOR CHAIN.

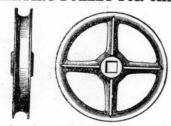

Placed at intervals between the Shaft and the Winding Engine, showing front and side view. These Pulleys are also suitable for wire or other rope.

SMALL FLAT ROPE SHEAVE.

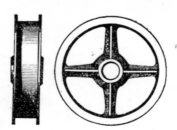

As used for the purposes of raising the produce of the Mine.

SMALL PULLEY OR SHEAVE.

FOR LIGHT PURPOSES.

PERRANARWORTHAL, CORNWALL,
AND
1 & 2 GREAT WINCHESTER STREET BUILDINGS, LONDON, E.C.

WEARING PARTS OF ENGINES, &c.,—*continued.*

MINING TOOLS.

WAGGON WHEEL.

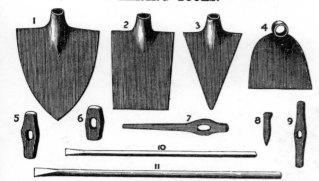

No.   DESCRIPTION:—

1.—Miners' shovel—sharp point, made of Iron or Steel.
2.—   ditto   square point.
3.—Bulling shovel, used for dressing ores.
4.—Spanish hoe.
5.—Sledge, used for breaking ores.
6.—Mallet, for beating borers 10 and 11.
7.—Miners' pick, with steeled head, for driving gad No. 8.
8.—Steel gad, for wedging off large pieces of stone.
9.—Cobbing hammer, used by girls in mines for breaking ores.
10 and 11.—Borers, for making holes in rocky ground to receive the charge of powder for blasting.

Williams' Perran Foundry Co. *keep a stock of these Tools, and can generally execute large export orders on short notice.*

These wheels have been made in large numbers by **Williams' Perran Foundry Co.** for waggons bringing down heavy loads of copper ore from the mountains in Chilli—and for severe and dangerous work in other mineral countries. Any special information and prices can be obtained on application.

CAST IRON BARROW WHEELS, WITH OR WITHOUT AXLES AND RUNNERS.

WROUGHT IRON BARROW WHEELS, WITH OR WITHOUT AXLES AND RUNNERS.

PULLEY BLOCKS, &c,

OF ALL SIZES, AND BEST PROVED SHORT LINK CHAIN.

PERRANARWORTHAL, CORNWALL,
AND
1 & 2 GREAT WINCHESTER STREET BUILDINGS, LONDON, E.C.

WILLIAMS' PERRAN FOUNDRY CO.

SUGAR MILLS

AND

MACHINERY.

Estimates given on application, according to the size required to deal with the produce of the Estates.

SUGAR PANS AND CASTINGS OF EVERY DESCRIPTION.

PERRANARWORTHAL, CORNWALL,
AND
1 & 2 GREAT WINCHESTER STREET BUILDINGS, LONDON, E.C.